OTHER BOOKS BY DAVID E. HARPER:

Wager The Wild Land:
A Paraplegic's Triumph over Tragedy

Vanquishing The Void:
A Paraplegic Discovers New Purpose

A Commander's Grace *for* Wounded Warriors

David E. Harper

Wasteland Press

www.wastelandpress.net
Shelbyville, KY USA

A Commander's Grace for Wounded Warriors
by David E. Harper

First Printing – September 2013
ISBN: 978-1-60047-904-5

Printed in the U.S.A.

0 1 2 3 4 5 6 7 8 9 10

To the U.S. military
servicemen and women
who have engaged America's enemies
and suffer injuries received
in their call of duty…

May God bless you!

I fled Him, down the nights and down the days;
I fled Him, down the arches of the years;
I fled Him, down the labyrinthine ways
Of my own mind; and in the mist of tears
I hid from Him, and under running laughter.
Up vistaed hopes, I sped;
And shot, precipitated,
Adown Titanic glooms of chasmèd fears,
From those strong Feet that followed, followed after.
But with unhurrying chase,
And unperturbèd pace,
Deliberate speed, majestic instancy,
They beat—and a Voice beat
More instant than the Feet—
"All things betray thee, who betrayest Me."

From the *Hound of Heaven*
By Francis Thompson (1859–1907)

Table Of Contents

"It's not our difficulties in life that define our character but rather how we face those difficulties when they come our way."

INTRODUCTION

Hellish forces of unimaginable chaos slammed against the side of the vehicle. Jaws of sinister basalt thrust at me through the windows ostensibly from all sides at once. Shattered glass filled the air. I spun uncontrollably amidst treasures that a few seconds earlier lay peacefully at my side. Light became dark. Then, in rebellion to an awful force outside of the car, screams of torn metal pierced the silence. The shell of a vehicle could no longer protect me. The world as I knew it forcibly changed forever.

Exposure to the raw force of gravity surged through the air. Like a bug in a hurricane, I suddenly found myself smashed against the windshield of life. In the aftermath, the hush of silence reigned in my ears despite the howl of the storm. I lay in a canyon crushed upon the rocks. As many can testify, life's brutal turns can send a soul into the gaping void of a bottomless abyss.

I held no real firm religious convictions when this accident occurred. For many years, I had listened to messages of salvation but stubbornly resisted any commitment to religion. Any problem facing me could be singularly negotiated with my fist or wit. A strong independent streak ran through me. My life's code bore this out: Real men didn't succumb to religion or help.

I believed God to be a crutch for the weak-hearted. With no regard for God or man, I took what I could from life and gave nothing in return. Lying in that canyon among the rocks in desperate agony, my soul craved anything that would liberate me from this horror. At last, I came to the end of myself. In desperation, I cried out to God for help.

I waited. I have withstood the furious blast of many storms since that fateful day in June of 1968, but none has been so cold or as terrifying. The demented wind mocked my plea. Its fury tormented the treetops until their moaning cry harmonized with mine against the rock walls before being silenced by the death shroud of snow-filled air.

When only the wind answered in reply, I cursed God, hoping if such a being existed those curses might anger Him enough to strike me dead. Did a god inhabit the world of men that a man could really cry out to for help?

The daylight hours waned like my anguished voice vanishing in the wind. Whistling its dire overture among the looming treetops, the wind bludgeoned their branches. A lacy curtain of snow spun around me as darkness engulfed the scene.

Thus began a night of unmeasured terror and agony. It hurled me into a vortex of despair as the gates of hell gaped wide to receive its own. Little did I know then, God indeed had heard, helping me to live in that darkest hour.

Since that night many years ago, this has become clear to me: God's power and His character are not constrained by human circumstances. God allows us to experience problems for our sake, to bring us into His strength when our power utterly fails us.

The Psalmist wrote:

> "I waited patiently for the LORD;
> And He inclined to me, And heard my
> cry. He also brought me up out of a
> horrible pit, Out of the miry clay, And
> set my feet upon a rock, And established
> my steps.
>
> He has put a new song in my
> mouth – Praise unto our God; Many will
> see it and fear, And will trust in the
> LORD." —Psalms 40.1

To say this event caused the worst pain and terror I have ever experienced understates my suffering. The intensity of the event threw me headlong into a turbulence of terror. My death screams howled through this murderous maelstrom for unconsciousness, anything that would release me

from the terrifying sirens from hell's wide open chasm
and the excruciating pain of consciousness.

Chapter One:

Whatever Your Circumstances, Don't Let Them Dictate Your Life

Life simply isn't fair! If it were, I wouldn't be in a wheelchair, and you wouldn't struggle with the circumstances in your life that bear down hard on you at times. Fair is that sense or belief that goodness and justice will prevail to make unwanted circumstances go away. It's a natural subliminal principle, which acts as a cornerstone to human thought underpinning all virtues.

Oftentimes, what's fair comes down to a matter of choices people make. Along with this belief comes a deep sense within the heart of each of us that tells us when something isn't right.

C.S. Lewis, one of the great intellectuals of the twentieth century, said that this standard that commonly exists within each of us is independent evidence that God exists. This standard is written in the heart of every human being.

For example, you can't tell if a line is straight unless there is a standard of measure to place alongside it. So it is with what's fair. Somehow all humans throughout history have acquired the same detailed standard of measure to know right from wrong. Of course the standard can be warped. I'll share more about that later.

We don't have a choice over many of the circumstances that we face in life. So, life is simply not fair. I had no choice whether I would be born a boy or a girl. For that matter, I didn't have a choice whether I would be born at all.

This is true for everyone reading this sentence right now. You had no choice over what family you were born into, whether you would be raised rich or poor, or what time and place you were born to live, did you? Did anyone ask you if you would mind being crippled, depressed, shamed, imprisoned or severely wounded in war? In every predicament, we often discover we simply had no say in the matter.

Oh, we may have had a choice that led to the mess we are now in, but once we are there, few of us have the choice to buy a ticket off of the path our life takes. Life simply is not fair. At one time or another, it seems as though we are a bug that has smashed against the windshield of life, which time sends careening sixty miles per hour toward us.

For many of us, things happen that direct our lives on a course we never would have chosen. A car accident leaves a child crippled for life. A bomb explosion on a battlefield leaves a fit young man maimed and blind. Divorced parents or sexual abuse sends a teenager on a path of shame and self destruction. Perhaps it's a broken relationship that destroys the ability to trust another person ever again, or maybe cancer strikes a young adult.

Suddenly, for whatever reasons, circumstances assail all human certainties, leaving us crushed beyond our imagination to recover. Have you ever had a bug-on-the-windshield experience? Have you ever felt like your whole life has ended in a train wreck or your ship to security, comfort, and happiness has smashed into a cliff, and you're left clinging to rocks, being lashed by waves and wondering what's next?

We know instinctively that this is not supposed to have happened to us. We look around and see injustice, affliction, and suffering and know, based on some instinctive standard written on our hearts, that the world is all messed up and this isn't how life is supposed to be.

So what about *God*? That same standard of right and wrong imprinted on the human heart also confronts us with the idea of a higher being out there somewhere. If God really exists, where is *He* when we need Him? Does He care? Stephen Hawking, a British theoretical physicist, cosmologist, and author, believes God set the laws of the universe in motion and worked Himself out of a job.

Some people may think that, by virtue of the fact that they speak English, were born wealthy, or simply have led a good life, they have a clear enough understanding of God. "Me and God have a good thing going," they think. Call it your god-in-a-box.

No matter who we are or what we have accomplished in life, God questions linger like an itch on our back that isn't easy to scratch and yet can't be ignored either. We grab at something to reach the itch, yet it's insufficient. The itch is still there. We may busy ourselves to forget about it for awhile, but when things heat up, the itch reemerges.

God questions are like that. We can grab for ideas and appeal to a god recreated in our own image, scorn those who do, disbelieve in God altogether, but no thinking human has ever been able to ignore the proposition.

When things go awry in our lives, God questions reemerge as we attempt to make sense of our situation. What can we do?

For many, after the initial shock over our life's misfortune abates, we become angry. It's a very normal human reaction. We look for someone to blame, someone to lash out at.

After discovering that this doesn't satisfy the agony they're suffering, some may push the blame off onto that god they imagined existed, right? After all, they reason—based on that visceral knowledge written on their hearts—he could have saved them from the dirty little hand that life has dealt them.

But in the end, blaming God or someone else doesn't really redeem life's unfair twist that has suddenly become personal, now does it? The problem still haunts. Guilt. Shame. Resentment. Anger.

Bitterness. Despair. Like an acid, misfortune can slowly destroy our souls from the inside out.

Friends begin to drop off like overripe apples from a tree. A few friends don't know what to say or they say the wrong things. Others can't possibly know what we're going through. Some, it seems, don't really give a rip. Even if they do care, often their advice seems empty. Sure, as if I'm going to just scrape myself off this windshield that's going sixty miles per hour down the highway. Right, thanks a lot.

Has the hope you once built your world upon vanished? Has it been replaced with a dull anguish in the depth of your soul? Does it throb like an open wound? In your loneliness, have you discovered a dark, foreboding, empty abyss you never knew existed? Oh, yes, and more. It threatens to consume you with every breath you take. Your soul begins to experience a hunger that cannot be satiated. It's most threatening when you're alone. Solitude becomes an enemy.

Recognize what I'm saying? You're not alone.

The abyss. It looms greatest at that point in life where we begin to ask ourselves, "What good is it to try to go on in life?" It grows when we ask, "Is there any good in life for me? Does God even care or moreover does He even exist?" The ache gets worse.

My dear reader, I know. I've been there. My heart grieves for those who experience great tragedy. I know the quest. If only we could find an escape. So, we search.

Sleep, drugs, alcohol or sensual distractions can temporarily steal us away from our living nightmare of unanswered questions or guilt, but then the next day's sun comes up and with it comes that empty, yawning solitude. Nothing seems to be able to fill it or drive it entirely away.

A wall of bitterness forms as a sort of self-preservation against falling headlong into this deeply personal abyss. What's frightening is that nobody seems to understand, right?

The pathway of anger, guilt, despair, and bitterness, however, tends to run only backward. It leaves us looking back at what might have been. We yearn for something beyond our grasp. Our heart aches. These feelings ultimately paralyze us. They serve little value toward recalibrating our lives to a new normal after devastating misfortune changes our life's course forever.

Burdened with a sense of helplessness, we're aimless, cast adrift in the vast ocean of life. With this, there's a sense that life has no purpose, and we have no vision.

Sadly, our future vision passes across a pain-filled horizon. Hope is not only beyond reach; it's abandoned us altogether. Life is simply not fair!

Do these statements sound familiar? If they do, my friend, perhaps this little book may help you find an anchor of hope. In the Bible, God makes this outlandish statement to a crushed and broken nation in Jeremiah 29:11,

> "For I know the thoughts that I think toward you, says the LORD, thoughts of peace and not of evil, to give you a future and a hope."

Hope? Yes, you say. Show me this hope. First I must say that true hope springs forth from faith and love and not from wishing for something that is otherwise unattainable. Contrary to other self-helps, Biblical hope comes from God.

This book doesn't offer simple answers to your problems. There are none. I've been in a wheelchair now for forty-five years, and the battles I face daily are real. I've stared into that abyss. Bitterness, hopelessness, worthlessness—yes, I've battled them all. Yet victory is available over calamity. Hope is there for all who are willing to risk their souls to find it.

How? It is through faith found in an unfailing companionship with the true God that created the universe. Not the god-in-a-box; rather, the God that created us. He fills the gaping abyss that haunts solitude. Then He pours concrete across the surface and stamps the word *LOVE* in the wet mud. On this, He'll build you a new foundation. In Him, there's contentment in unfailing hope.

God offers freedom from the bondage of misfortune in our lives. Bondage creates a sense of emptiness that can be combated with the weapon of faith: faith that Someone cares, and hope that He will right all things in the end. You too can know this hope built on a love worth living for. The knowledge of true hope breaks the chains of those held captive in despair.

Chapter Two:
The Search for Hope

There is a death worse than dying. That is the death of hope. My heart aches for those who live without hope. I know personally that hope is an asset to those who find it. Those who do will have courage to go on with life, despite life's terrible predicaments.

Armed with hope, people can ride out and surmount difficulties, tenaciously hanging on to their own remaining potential in life. Hope brings the courage to look at whatever meager strand of life we have with a sense of eager anticipation. Hope looks forward instead of backward.

"HOPE *is like* *a* FRAGRANT RAIN *that falls on a* PARCHED DESERT."

It brings an end to struggles and ushers in new life.

No matter what our situation in life, no matter how useless we've become, there is One who cares about what we're going through right this minute. Jesus Christ cares deeply. He not only cares, He also provides a way out of hopelessness. In replacing hopelessness, He offers each of us the anticipation of a vital purpose in a vast plan that leads forward to the kingdom of God.

Jesus cares about the dreadful anguish that people endure. Jesus wept at the tomb of Lazarus, even though He knew He would raise him from the grave.

In love that only God can provide, He demonstrated compassion for those grieving at that gravesite.

To those suffering from the effects of demoralizing distress, He offers peace. He holds out nail-pierced hands for those tortured by struggle. The Man of Sorrows offers comfort to those paralyzed by agonizing grief. To those who live in pain, He offers hope from a face beaten and scarred by thorns. The man of mercy draws those riddled with shame and guilt to His spear-scarred side. Strength and courage are available to those who earnestly seek it from the Source of life.

The immortal walking out the death sentence offers mortals freedom from bondage.

Through Jesus Christ, God offers you the brave heart of courage that will eventually breathe satisfying joy into a deadened life. The courage to press on, to struggle against confining circumstances is one of the virtues of living. It's where you can discover life's true potential.

The Bible tells us that the ungodly promise liberty while they themselves are slaves to corruption. Think about it for a moment. Are you enslaved by the corrupt ideas of people that have prevailed upon your life and violated your trust?

Honestly, what keeps you from moving forward with your life? Is it the brick wall that you've been slammed up against? Have you been betrayed by a spouse or close friend who has misplaced your trust? Perhaps an accident has crippled or disfigured you. Or perhaps a prison steals your freedom. Not all prisons have walls, you know.

Are you one of those who suddenly discovered that religion has let you down? If you are, stay with me. What I'll share with you is not religion. Religion is merely man's attempt to reach God.

There is value in suffering if you don't allow it to enslave you. A season of suffering is worth that moment of glory when the light finally shines through.

"*It is not* OUR PROBLEMS *that eventually define* OUR CHARACTER; *it is the way we* FACE *those problems.*"

Yes, everybody's limits are different. Some people do indeed endure much greater hardships than others could even imagine. Some burdens are much greater to bear than others. You may be one of those people.

Everyone, however, can live within their remaining potential by discovering the strength and power of God in Jesus Christ. Whatever suffocating darkness looms in life, confusing every thought, Christ is there. Call out to Him. He hears even the weakest prayer and will respond but not always in the way we expect Him to answer.

God offers strength to stick-it-out through life's disappointments with an eager anticipation toward a future and hope. Allowing God to work in our struggles gives us hope to overcome difficulties and fill the abyss in our souls.

God created every human being in His image. He has crowned mankind with a special kind of majesty. In that majesty, God offers a divine purpose in this life. This purpose is discovered by moving beyond crippling changes that consume our lives. These changes can be like doors. Every door in life is like an entrance into a new experience. So sit back and take a journey with me.

As a youngster, I loved to be alone in places void of people. I could spend hours and days hunting and wandering alone. My heart's desire drove me to aim for a job working outdoors. This was the life I loved. I gathered strength and energy from the outdoors away from people. People always seemed to bring me turmoil.

Then at the age of sixteen, after a weeklong trip in the mountains, the car I traveled in slid off the road and plunged over a steep precipice into a canyon below. Part way down the mountain, the centrifugal forces operating on the car rolling down the mountain ejected me through the window, hurling me through the air onto the boulders strewn across the canyon.

I suffered a broken arm, a compound fractured leg, and many broken ribs. When I landed on the rocks, the trajectory snapped my backbone. A severe head concussion, fractured skull, and severe lacerations across my face left me mercifully unconscious. Cold weather settled over the scene as snow swirled through the rocks.

Draped between two large rocks like a soaked beach towel, I lay partially in water. With consciousness came violent waves of shaking from chill, sending excruciation pain throughout my broken body grinding bone-on-bone. For several hours, my life drained from me sending me into deeper shock. I was dying.

I knew of the religious stuff preached in churches, but I did not believe it. To my thinking, religion was for women and children. There was no problem that I could not face alone.

My belief then was that real men didn't succumb to religion. God, I believed, was a crutch for the weak-hearted. With no regard for God or man, religion offered nothing from the harsh realities of life.

I did all I could to marginalize God. To me, God was like people. I thought He would bring only trouble. My god resembled a bottle with contents that had more control over my life than anything else. When drink came available, I drowned myself into a drunken stupor.

Chapter Three:
God is Not Constrained by Your Circumstances

Lying in that canyon, desperate for relief from the waves of pain racking through my body, I found myself crying out to God for help. But because I didn't have any reference point about who or what God might be, I didn't know if He even heard me.

A miracle would have been nice. *Just let me stand on my own two feet, God,* I thought to myself. After that I won't need your help, and you can go back into your box. My voice, fading into the wind, seemed to disappear among the mountains like the snow that enshrouded the air. Beyond the howling wind, silence reigned with no tangible help from God.

After that, I began to curse God in hopes that, if He really existed, I might anger Him enough to strike me dead. Still the empty silence reigned. Although a friend who had survived the ordeal nearly uninjured tried to help me, I felt alone. *To cry out for help in death's icy grip and think that no one has the power to help terrifies the soul, leaving only anguished torment.*

This is the state of those that do not know the true God as they grapple with overwhelming challenges in life. To many people, normal challenges up to that bug-on-a-windshield moment in life seem as though they can be battled through with strength and grit and some occasional help from their god-in-a-box.

Normally we do not give our god-in-a-box, stored for rainy days, a second thought. But then one day, something beyond our worst nightmare occurs in our lives. They are viscously ripped apart. The world turns upside down. When we turn to the god neatly packaged in a box, we find the box empty.

The real God, whoever we think He might be, has no relevance to life until that moment. Fleetingly, thoughts of Him as "the man upstairs" linger, in case we need help. Others acknowledge Him enough that they may say, "Me and God? We're good, ya know." At the moment of catastrophe, though, any of those attitudes will not help.

For me, I felt alone in that canyon. The day faded and night came early because of the storm. Little did I know, however, that God indeed heard my cry for help.

You see, even though my situation seemed hopeless, God came alongside me. He heard and helped, because, as an image bearer of God, He cared about me just like He cares about you. He hears even when you think He does not. You see, God's power and character are not constrained by human circumstances. His concern for humanity does not fall short because of human attitude toward Him.

It is written in the Bible that *God demonstrates His own love toward us, in that while we were still sinners, Christ died for us.* Imagine that! While we were His enemy, Jesus Christ died for us.

I convey this testimony for one reason: That I may speak boldly to a particular person reading this little booklet right now. Perhaps that person is you, my dear friend. Do you reside in a canyon of despair racked by pain brought on by circumstances you are a victim to suffer by no plan of your own?

Maybe you are a teenager searching for answers about God. Perhaps you are a battle-scarred soldier returned home, defeated in war, and abandoned by the government you stood up to serve. Prisoner? Do you languish in a prison because of some mistake in your past? Or perhaps you're marginally religious, grappling with the heavy burdens that life has laid on you, but your real god is addiction.

Maybe you are like me in that canyon, not believing all of this God stuff: Are you someone who has heard that God cares for you, but you have chosen not to believe it? That's okay. God looks for an honest heart that willingly asks, "Are you really there, God?"

Well, God has caused you to read to this point here to learn an old, old story.

> "JESUS *loves* YOU."

God gives us problems for our sake, to bring us into His power and strength when our power utterly fails us.

Jesus says in Matthew 11.28, "Come to Me, all you who labor and are heavy laden, and I will give you rest."

In South America, I saw a man struggling under a bundle of firewood twice his size. Bent down under the bulging weight, he looked worn out. Oh, this was not just a one-time load. The glazed look on his face revealed how routine his load had become. His bare and calloused feet betrayed the years of his struggle.

Perhaps the path he had walked daily felt like a treadmill going nowhere. Then again, maybe he had given up caring.

Do you feel worn out by the load you carry? Is it a load you alone carry? Once you stood tall, feeling fit. You had the best the world could offer. Your job spelled *success.*

Then something beat you down. Your colleagues kicked the career ladder out from under you. The unthinkable happened. Was it broken trust? Perhaps a war injury? Maybe you were stricken by some rare disease that only one out of a million people have ever even heard of. Whatever the case may be, today you're worn out. Your calloused heart bears witness to you alone of the weight of your unique burden. Secretly in your heart, do you groan, "Why me?"

Jesus knew pain as we know pain. Our problems are the silken cord that God often uses to bring us to Himself and into the rest He alone offers. The God who is concerned with the details of the universe has not overlooked you. *God loves you!* He offers you the *rest* that you just never seem to be able to find. Not just a nap kind of rest but true rest for the depth of your soul.

In that canyon, I suffered the most immense pain I have ever known. By and by, a person driving by noticed skid marks in the snow, leaving the road and disappearing into the abyss. He could not reach us but went for help. In the days prior to cell phones, hours went by before help arrived. More hours passed as my rescuers struggled down the face of the escarpment to reach me.

In the hospital, they told my loved ones that I would not live through the night. But God had a purpose for me just as He does for you.

"GOD *desires* GOOD *for* YOU."

No matter what men may say. No matter what that dark voice inside your head mocks you into believing. God wants to fill the desires of our hearts if we are willing to place our faith in Him. However, the good He shares with us is not always what we may consider good.

I lived through a hell on Earth that night and many more to come. Recovery was slow. For two months, I lay absolutely flat in the hospital bed. A couple of more months went by as I learned to sit up and move around in a wheelchair.

After five months in the hospital, I went home to a meager existence. A once-strong body now adorned a wheelchair in a crumpled heap. I had lost the strength of my youth. I had lost the ability to be independent. No longer could I stand straight as a man should stand. I viewed myself as a half-man, and a half-man is no man. These were insurmountable obstacles to me. I had so many obstacles to overcome

I wondered if life was worth living. The burden became unbearable.

Initially, I thought God had cursed me for my wicked life. I thought that even if I wanted to come to God, He was fed up with me, that He was through with trying to save me. I thought that a crippled broken body was the result of my irreverence toward God, and I deserved what I had gotten.

It felt as though my life no longer had any value. For many months, I could not go to the mountains and be alone. I thought my life had come to a dead-end road of misery. But this was the furthest thing from the truth. An unseen Providence protected and guided me every step of the way.

This was only a turn in my life's road. This tragedy became a beginning point, although it did not seem so at the time. Over the ensuing years, I discovered an interesting fact. Obstacles to me are vehicles to God.

Chapter Four:

Obstacles are Vehicles to God

At the core of living, all humans are universally unprepared for ruinous setbacks. No one is prepared to react to unexpected devastation. The point is not *that* we react to the unexpected. Animals react to the unexpected. It is a matter of *how* and *why* we react that makes the difference and distinguishes humans from animals.

Most of us have been faced with a situation and then later thought, 'I could have done this' or 'I should have said that.' That is precisely my point. It's not our ordinary reaction to the common experiences in life that gives us significance. It's the unexpected and how we handle it. This not only makes the difference in our lives but also affects the lives of those around us. Our passage on Earth is not just for our sake; it is also for the sake of those within our sphere of influence. As John Donne wrote, "No man is an island, entire of itself."

We're all on this ship called Earth together. Although we enjoy many amenities, we don't need to look far to realize that something has gone terribly wrong on this planet. Bad things happen to good people. What seems fair doesn't always pan out that way.

The Bible possesses an accurate description of the world. Instinctively, we know that we should have dominion over the earth. But the Bible teaches that humanity has an adversary named Satan, who has invaded our world, seizing dominion and wreaking havoc among us. Not only has he invaded our world, but he has also darkened our heart and minds. This prevents us from seeing truth as God sees it.

On the other hand, the Bible teaches that God is the most powerful being in the universe. Vast beyond all measure, He knows all things, even the secret things of the human heart. There is no place that we can go where God is not there. Our sun is like a child's ball to Him. He spoke and all things were created. Since everything that exists has been created by God, He owns everything.

Perfect, holy, and righteous God cannot tolerate wrongdoing. He has never violated His principles. He will not allow anyone to stand before Him who has violated His law. He exists eternally separated from any form of lawlessness or unrighteousness.

His principles are eternal laws. You could call them the laws of nature, or the laws we instinctively have written on each of our hearts. They are laws that allow us to determine what is fair and unfair, righteous and unrighteous. Leaving the Bible out of it for a moment, we still know that the world is messed up because of wrongful instincts shared among humans throughout history.

Of all things in the universe, everything operates in perfect harmony with God's laws except mankind. All the stars in the universe operate according to cosmic laws such as the law of gravity, the law of motion, and the law of electromagnetism. Light functions perfectly according to the law of light. All living organisms on Earth function according to the law of biology. Trees grow and animals live by the behavioral laws ordained for them by God. An exception to all of this is man. Of all material things in the universe, only man alone can *choose* to behave outside of the laws God established for mankind's behavior. We call these moral laws.

God created mankind in His image to be free like Him, even free from Him if we choose. And so we have chosen, as a result of Satan's influence, to deny God. Even so, throughout history and across all cultures, the same basic law written in the human heart has not changed. Murder has always been murder, taking another man's wife has always been wrong, as well as lying and stealing.

Individuals may choose to call wrong right or right wrong. Criminals may attempt to justify themselves, but there is never any doubt in a victim's mind that they have been unjustly treated.

A person or a society may choose to operate outside the laws of nature, but in time others will bring their actions to justice based on the moral law written in the heart of mankind.

History repeatedly testifies to this maxim. Attila the Hun, decadent Rome, Genghis Khan, Nazi Germany, Pol Pot's Khmer Rouge, and Stalin's Russia all have been judged in the heart of future generations for deviations from the laws of nature and nature's God. They all thought they were in the right. All believed they could justify their actions.

Any breach of God's law perpetrated in our hearts or toward another human being makes us guilty in

God's court. This guilt proclaims us unrighteous and automatically separates us from God. We may think we are good and, by human standards, we may be. But in God's court, we will not be judged by human standards. We will be judged by His law and not the god-in-a-box's standard that we have created in our minds.

You would be better off confronting an angry grizzly bear and demanding that the bear change its ways to your idea of a bear-in-a-box than to stand before God and demand that He must calm down because His ways are not yours.

All of us have at one time or another lied, cheated, betrayed a trust, and consequently sinned by breaking God's law. It takes only one time from the cradle to the grave to make us guilty. Humanity is guilty, as is each of us individually before God. We've taken our freedom and used it to reject the moral laws God placed within us. The penalty for our guilt is death.

All of us stand condemned and subject to eternal separation from God in Hell.

God's anger burns against all unrighteousness. In our sin, we've affronted God, declaring ourselves, by our actions, to be His enemy. We can't simply change our habits and be okay from now on. He demands a penalty be paid in blood for even the smallest breach of His law. Only a person who has never committed the slightest wrong and has been perfect their entire life can stand before God. Who can make such a claim? I certainly cannot. Can you?

Suppose a man performs a heinous crime against you or your family and goes to trial. There he declares to the judge that he has seen the error of his way and has decided to go straight. It is obvious to everyone in the courtroom that this man is manifestly sincere. His reform makes him a good man, right?

Would you expect that judge to say, "Since you have reformed yourself and proved your sincerity, I will forgive your crime"? No. We all know the person must still pay a price for their crime and be punished for breaking human law.

Because God is just, He has also assigned prophets to write down all His laws. Over the course of human history, God preserved His laws so that everyone throughout history would have a chance to know what is expected of us. God's laws for mankind are summed up in the Ten Commandments. So, these laws are written not only on each of our hearts but also in stone.

> "GOD *is a righteous* JUDGE."

He has far more integrity than any human judge. How much more then will a sovereign God in His righteousness demand that punishment be served for breaking *His* law? Don't think for a minute that because He is loving and kind, He will graciously excuse sin. Yes, He loves; but remember He is also *just.*

If you're a man, you can't just man-up. Ladies, although your friends all agree you're a good person, their claim that you're a virtuous woman will not stand in God's court. None of us can claim we're good enough. Neither can we determine to reform our behavior and go straight, or become more spiritual and expect God to forgive our past violations of His law. By our collective lawlessness, humanity has allowed Satan to have his way on Earth—and at one time or another in each of our own hearts. The bottom line is that each of us has chosen to break God's laws, and sin.

Equally, Satan hates all image bearers of God because he hates God. He is the source of many of the afflictions that enslave us physically and spiritually. However, Satan wants us to blame our troubles on God, continue in our trespasses against God, and self destruct. In this way, Satan keeps the dominion intact. So we're trapped in sin, enslaved by circumstances, and liable before God who demands punishment for violating His laws.

But wait. There is good news.

Chapter Five:
Self-evident Truth

Now, all you may want is for an injury to heal, a relationship to mend, cancer to go away, or your life to return to normal. All this may seem irrelevant to your particular situation. I fully understand. After forty-five years in a wheelchair, I truly empathize with the longing of your heart. If you've read this far, though, then some of these truths must resonate with you. Perhaps you're beginning to see why life isn't going your way.

Remember we don't always have those choices available to us. What I hope you will see is that the afflictions we suffer physically are compounded spiritually. To ignore the spiritual aspect of our suffering sets the problem aside only to be dealt with later. Before we can hope that God will meet our physical needs, we must back up a few steps and address our spiritual condition.

We are created with freedom, freedom to choose our destiny in some circumstances. Looking to God for help is one area of our lives where we do have a choice. It is where some of us will discover our personal need for God. On one hand, we're left with the dilemma of dealing with physical afflictions we are unable to change. On the other hand, we're given a choice of how we face those afflictions.

We can receive nothing, of course, by spiraling downward in despair and closing out God. Some may try to overcome their afflictions on their own by getting tough. To those I say good luck. Come back twenty years from now if you are still alive and let me know how it turned out for you. Others may seek self-help (some, however, only as a last resort). That is certainly a choice.

Before you set this book aside, at least discover the good news. As I said, God loves you. God has a gift tailored just for you. However, because we're at war with God, we often don't want to hear about this gift until we have one of those bug-on-the-windshield experiences.

It is at this juncture in life's road that God meets us. Despite our offense to His law, God has compassion for those who have come to the end of themselves and cried out to Him for help. To those who accept His gift, He offers courage and strength to enjoy the remaining potential that is left in life.

Only God could meet His own requirement of being perfect and without sin. So, God came to Earth and became a man who lived a perfect life. This God-man is Jesus Christ. He alone met God's perfect and sinless requirements to stand before God.

Then with unbounded courage and love for mankind, He took on our condemnation. Jesus submitted to the contempt, ridicule, and hostility of mankind whom He had come to save. This He endured through a torturous death by Roman crucifixion. By so doing, He shed his own blood and paid the price of our guilt brought on by our lawlessness.

Jesus suffered anguish beyond measure for all who would accept that He paid the price for our disobedience of God's laws. Jesus was passionate about this, and He broke the bonds of hell for you, my friend.

He did this with the joy of knowing that His sacrifice would allow us to spend eternity with God. In three days, He rose from the grave to prove that He alone had the power and authority to pay the price of man's sins, overcome death, and escape Hell. He literally died and went to Hell for us, joyously paying our price for breaking God's laws. As such, Jesus is alive today with God in bodily form.

All we have to do is humbly acknowledge our condemnation before God, repent of our sins (our lawlessness before God's court), deeply appreciate the grace Jesus offered when He took our punishment by dying and going to hell in our place, and believe that He alone possesses the authority over death by rising from the grave. The result will be to live our lives in appreciation for what Jesus Christ has done for us. That is the good news. We can assume the righteousness of God, avoid His wrath, and one day live with Him in heaven because of Jesus Christ.

We can't do this, though, until we suffer the anguish of acknowledging that we've done something terribly wrong, and the penalty of sin will fall on us with a fury.

In Mark Twain's book, *Tom Sawyer*, Tom surprises Becky Thatcher sneaking a look at Mr. Dobbins, their schoolmaster's private book. Startled, Becky snatches at the book to thrust it into the drawer but, in doing so, tears a page. Later, Mr. Dobbins discovers the torn page. In scornful wrath, he eyes the room and, in turn, begins to ask each student if he or she did it. Terror reigns as the inquisition mounts. Becky looks like a scared rabbit as Mr. Dobbins approaches her desk. "Rebecca Thatcher...." Instantly Tom springs to his feet shouting, "I done it!" for which he suffers an unmitigated flogging by the schoolmaster.

Few will deny the honor of Tom's action. Fewer still would deny the gift if they endured the anguish that Becky suffered. Now if a most unlikely hero might take the fall for something he didn't do, why is it so hard to believe that a loving, honorable God could be less noble? Yet many reject the gift that God offers through Jesus Christ. Why? Because Satan has darkened human hearts and blinded us from truth.

When Jesus hung on a cross, dying for our sins, two thieves hung on crosses on each side of Him. One scorned Jesus; the other recognized the power of God in Him. This one called out to the other thief,

> "Do you not even fear God, seeing you are under the same condemnation? And we indeed justly, for we receive the due

> reward of our deeds; but this Man has done nothing wrong."
>
> Then he said to Jesus, "Lord, remember me when You come into Your kingdom." And Jesus said to him, "Assuredly, I say to you, today you will be with Me in Paradise." —Luke 23:40–43

As it happened in those days of old, a few today will recognize that Jesus Christ is the power of God's light shining in the darkness. Some with broken spirits will say to Jesus, "Lord, remember me!"

If you are among those who wish to receive the blessing of Jesus Christ's sacrificial gift from God, you can ask Him right now. I won't coax you in a prayer like so many others may do. If you really understand and anguish over condemnation for breaking God's law and know you'll suffer the fury of God's wrath when you die, then bow your head and speak to God about it right now. Let Him know how repentant you are for your sins and how grateful you are for Jesus's death on that cross for you. Seek God's grace; it's mercifully offered to you from nail-pierced hands. This is not a religion. This is God relating to you personally.

Next get a Bible and start reading it. I recommend you start by reading the Book of John. As you read, pray for understanding and pray that God will send someone into your life to disciple you in a deeper knowledge of what you're reading from God's word.

However, if you don't completely understand what's being shared with you here and sense no anguish over God's fury toward you, then please do not make a prayer of confession. It would be better for you to live under condemnation than to believe that you've been saved from God's wrath when you really haven't.

Eventually, you'll fall away from your belief while thinking you tried this *Jesus thing* and it didn't work. It won't be that it didn't work, but rather that your spirit hadn't been broken in anguish by the realization of God's great love for you and yet His full wrath on your eternal soul for breaking His law. Yours would be a life lived under an awful veil of deception.

If this only partially resonates with you, though, ask God to explain it to you. He is gracious toward people willing to seek Him through His written Word. Find a Bible and start reading. He will guide you in the knowledge He has to offer you. The decision to follow Jesus is the most important decision you'll ever make. It is not one to be made with little forethought.

Personally, I didn't immediately turn my life over to God through Jesus Christ. Even though some of it resonated with me, I thought I was good enough and my sins were more of a joke on the world. Viewing sin as a crime against God and deserving of His wrath seemed distant, rather than personal. As such, a need for the gift of God offered in the sacrificial work of Jesus on the cross felt irrelevant. I tried to make a go at overcoming my difficulties by occasionally going to church and posing as a Christian. In reality, I was nothing more than an imposter—a poser—fooling only myself. Yet something within me sincerely sought the truth.

As I sought to understand God, He gradually showed me the truth about the power He held in my life. This truth would one day come to me.

In the meantime, He showed me that from my brokenness, I had potential to relate to others suffering from afflictions. I didn't understand how that potential could affect others, but I found a measure of peace in knowing that it did. That peace was incomplete, however, without putting Jesus Christ into the equation.

This knowledge didn't come through clever insight on my part or strength rising from within me. It came entirely from God. He gave me the ability to encourage others to overcome their own troubles. My personal obstacles became vehicles to bring courage to others.

God seemed to ask, "Can a wheelchair weaken *My* arm?" Defeats, failures, and mistakes are inevitable parts of our lives. God wants to help us deal with problems and turmoil so that we can fulfill His purpose in our lives. Before He helps us, though, we must understand our need to repent of the sins we've committed by breaking God's law.

As babies, we learned to crawl and walk and overcome our obstacles. Why do so many people believe that overcoming obstacles will not continue throughout their life? Don't surrender to your problems too easily.

Calvary was a great tragedy for the men watching Jesus Christ being crucified on a cross and yet a great victory for God and ultimately mankind. We need to look beyond our obstacles and not allow them to define our lives.

Defining our lives by previous tragedy is living in the past. When we do this, the past controls us. Relinquish life's crushing circumstances to Jesus

Christ. Live for the future destiny you have in Him. To those who search for their destiny in Jesus Christ, God will provide the power to overcome terrible tragedies.

God offers strength and hope to put obstacles behind us. He said to Moses thousands of years ago in the Sinai Desert when the Israelites cried for meat, "Has the LORD's arm been shortened? Now you shall see whether what I say will happen to you or not." —Numbers 11:23. Afterward He brought in enough quail to feed six hundred thousand men, plus women and children, for a month.

Will knowing and accepting these things make all your pain and sorrow go away? No. Will you gain the strength and courage to go on with your life despite your afflictions? Yes. But greater yet, you will receive the righteousness of God through Jesus Christ and be assured that one day God will make all your misery vanish.

Don't be deceived into thinking that because God hasn't eliminated suffering and ended Satan's dominion in this world, He can't. God can and He will. The Apostle Paul wrote in the Book of Romans 8:18,

> "For I consider that the sufferings of this present time are not worthy to be compared with the glory which shall be revealed in us."

Ultimately, you'll receive the peace of God that passes all understanding. This peace will keep your heart and mind in Christ Jesus through whatever suffering or injustice you may grapple with.

Chapter Six:

In Our Weakness We Gain His Strength

Obviously, the things described here do not lead to an abundant life filled with fun and happiness. If you want fun-filled happiness, you won't find it here. Leave that for the fair or carnival.

What the Gospel message offers is something far nobler than fun and happiness. It offers an understanding of why bad things happen to good people. Furthermore it provides a peace that will keep our hearts and minds anchored on something bigger than ourselves. Joy emanates from this knowledge.

God's purpose and plan in your life comes from a love so great that the God-man Jesus took the punishment for your sins as a gift to you. The Bible says that Jesus endured the cross for the joy of your salvation set before Him.

If you accept this gift, God will fill you with the joy and courage of Jesus Christ to live in His strength. Through the courage and strength that God offers in Christ, which He demonstrated on the cross, God can work out things in your life that are not humanly possible. Your life may never be what you expected, but it will be much more abundant in the purpose God has in store for you.

Jesus said that Satan comes to kill, steal, and destroy, but that He comes to give you abundant life. Some people's lives have been racked by trouble and circumstances brought about by destructive forces from Satan who desires only to mess up and destroy their lives. Does this resonate with familiarity? Does it seem as though everything in life works against your efforts to succeed?

Satan is a real enemy actively working in this world to destroy the goodness God desires for us. Rather than seeing God as a cosmic masochist, recognize that your real enemy is Satan. God offers a real solution through the power of Jesus to stand firm and fight against the struggles we experience. By accepting God's solution, we become perfected in that which He created us to be.

Jesus calls us to something more gallant, dignified, powerful, and just than mere fun and happiness. Through Him we are a new creation, no longer in bondage to sin and the destructive forces Satan uses against us. As part of this new creation, we receive a new purpose in life. Again, this is not to say our struggles and afflictions will automatically go away. Mine have never left me, and yours may not either. However, by following Jesus, those struggles will be secondary to what God calls you to do.

It must be very clearly understood at this point that if someone accepts the gift of God's grace only to work out some personal plan, it will result only in disappointment.

> "GOD'S GRACE *is extended for* GOD'S PURPOSE *and* GOD'S PLAN."

Does this make Him seem like an egotist? Well, ask yourself, who more than the perfect loving God who laid His life down for yours deserves such honor? When this becomes clearly understood, then God's blessings will flow through Christ's sacrifice, strengthening life with hope. Not only hope, but through tribulation, we discover the courage to persevere, and with perseverance, strength of character emerges. Finally, grounded in these things, we learn to find joy and added hope in the glory of God, which does not disappoint.

In the Rocky Mountain West where I live, there's a legend told of an old warrior chief. Among the tribes, his name resonated for battles they had fought, people who'd prospered, lives saved, and greatness gained by his nation. To one of his three sons, he desired to pass along his mighty name. This son would become destined to continue in his footsteps and lead his people. First, however, this son needed prove himself worthy.

So the old chief gathered his three sons. He instructed them to take a long journey to a distant region where few men have ever traveled. In this faraway land, a mountain would rise so high that its summit was rarely seen. Many people claimed that to go there meant certain death. This battle-scarred chief instructed his sons to climb this mountain to gain vision of the world around them. The son with the most vision would be deemed most worthy.

The chief's most self-centered son thought his father's plan was foolish. This son felt he had more important things to do in his life. He reasoned that if he returned first, his father would give him charge over all things, making him very popular. So he didn't climb the mountain very high. After many weeks, this son returned.

The old chief asked his son, "What did you see?"

His son replied, "I climbed for many days. Food ran short and I suffered greatly, but I finally came to an enormous spring. It's from this place that the mighty river from which all life has its source flows into our country."

"It is well, my son," said the old chief.

Weeks later, another son returned. This son desired to accomplish his father's instruction fully but lacked perseverance. Ultimately, when the wind and rain raged against him, he faltered. At a high elevation, cold set in, but he pressed on until he passed beyond the timberline. This landscape became strange and unfamiliar. There he began to think of his home and friends.

On his return, the old chief asked this son, "What did you see?"

"I traveled with little to eat until the timber grew very thick. My horse could no longer climb the slopes. I fought through ancient trees, some blown down by the wind and crisscrossed on the ground. Finally, I broke through the trees and entered a rocky land too cold for the trees to grow. No man could live there. The clouds obstructed my view, and I came to believe I could be more valuable here than I could be stumbling about in clouds.

"It is well, my son," said the old chief.

The last son wondered about what his father had sent him out to discover. He pondered over this thought as he pressed to mountain's great height. The forest tore his buckskins, yet he pressed onward. The rock above the tree line tore his hands and feet, yet he never faltered from the belief that his father had a purpose for sending him there. He suffered great afflictions until he reached his goal.

Much time passed before the last son returned. Likewise the old chief asked this son, "What did you see?"

"Father, I climbed past the source of the great river that brings life. I stumbled through cloudy slopes strewn with rocks where my vision failed to guide me, yet I pressed onward. Dangers, toil, and afflictions wore me down, and I grew very weak and tired, nearly giving up and wishing to be home, but then I saw it. From the mountain's windswept summit, the clouds parted, and I viewed a great body of water that stretched from horizon to horizon. I discovered something exists that dwarfs all I knew to be true."

With insight born on experience, the old chief said to this son, "It is well that you did this, for I know that only one who completed the ascent to the mountain's great height could know of such a thing as this. For your vision, your perseverance to endure hardships, your selflessness, and faithfulness to my instruction, I give to you alone all I have."

God, our Father in heaven, desires to give us much more than we could ever imagine. Such a gift, however, does not come without consideration. Are you willing to accept that the only path to heaven lies in accepting Jesus Christ's suffering and death as your personal ransom from sin and God's wrath? Are you willing to endure suffering and injustice and climb heights unseen by others and live for something greater than yourself? It will be hard but rewarding.

A truth resonates within the Indian legend. The Bible says that weakness can strengthen us if we persevere, perseverance builds character, character brings hope, and hope sustains through the suffering and injustices we see in the world around us.

As an athlete trains in a gym day after day, his muscles feel sore and weak. But training is necessary for him to build up his muscles. If he says, "I will take it easy and not work out so hard," he will not be fit for the race. But if he pursues his training and strains his muscles to the point of weakness, he will eventually become strong.

Here lies an eternal principle: Strength comes from weakness. Those that quit and hang their lives on a wall generally do not achieve much in life. If we are going to achieve God's goals in our lives, it's not going to be easy. It will require weakness and weariness. When we understand that in weakness God manifests His strength, we'll be encouraged to continue on course. The Apostle Paul once said,

> "For my strength is made perfect in weakness. Therefore Most gladly will I rather boast in my infirmities, that the power of Christ may rest upon me." —2 Corinthians 12:9

This encourages us to go forward in the face of difficulties, obstacles, and discouragement.

"ENDURING WEAKNESS *builds character in* OUR LIVES."

Simply stated, to overcome is victory.

Personally, this realization came to me one night after my life fell into utter despair. God allowed me to come to a point where I desperately longed for something to fill the void within my soul. The truth finally came to me. All of my cleverness, all of my wit, all of my knowledge, and all of the prestige I had gathered failed to fill my soul with peace. All of my self help and independence were of no avail.

My job was not going well, and my marriage was faltering under life's storms. I was not having the relationship I wanted with my son. Two of my colleagues at work were setting me up for a fall. At these crossroads, He showed me a third truth. He showed me two roads. One road was my life as I wanted to live it. The other road was the life He had planned for me. He gave me a choice of which one to follow. My life the way I wanted it was broken. Something was very wrong. I read in the Bible that the ways of men without God lead to a bitter end.

Finally, I realized my capacity for understanding life's greater complexities was limited by my own fabricated and simplistic look at life and my god-in-a-box. Like the first two Indian sons, my unwillingness to go higher thwarted my capacity for future hope.

We can't live a life well enough for God. That is why He did it for us through Jesus Christ on the cross of a windswept rocky knoll. All He asks is that we surrender our burdens and make Jesus Christ Lord of our lives. At that point, we'll discover the top of the mountain in our own life, viewing with vision a God-like perspective. On that summit, we'll discover that there's a difference between simply believing in Christ and giving Him Lordship over our lives.

As with the third Indian son, in amazement you'll discover a world that dwarfs everything that exists as you know it.

> "GOD *has* A PLAN *for* YOUR LIFE."

Are you willing to risk everything to discover what's at the top of the mountain? You see, for everyone that surrenders their life to God, He will assign a heavenly purpose. That purpose is oriented to a kingdom, God's kingdom, something greater than anything that anyone can imagine. By living out that purpose, God will do through you great and marvelous things, accomplishments worth dying for. God wants us to show people from all over the world the truth in Jesus Christ.

Chapter Seven:

What are You Willing to Risk?

Remember when we started down through the pages of this little book together, we talked about choices? Well, we may not have a choice to get off the path we're traveling down, but we do have a choice of how we'll face the journey ahead of us.

I can't give you a satisfactory reason for why events or circumstances in your life or for someone you know have gone the way they have. The world in Satan's power is not as God intended it to be. Does this mean that God doesn't care about our suffering? By no means; He can and will one day correct all things. However, in the meantime, I can say emphatically that God provides a way to overcome and endure life's enormous challenges.

Whatever your particular circumstance may be, you have a choice. You may sit on the couch in front of the television set or computer and feel helpless.

However, humans are exceptional. Human beings are the crowns among all of God's creation. Thus we have been granted the greatest freedom of choice. Remember that of all the laws of nature, man alone has the ability to choose to disobey the laws of his own nature.

When people cross over these laws, order and purpose in life begin to unravel. Someone has violated the law of human nature given to us by God. In their thinking, what they did may be *no big deal.* In the other person's mind, though, it is a major thing.

Yes, it is true that a person or a society as a whole may say, "I see no problem with lying or cheating," and another may say, "I have no problem with taking another man's wife." The human conscience is easily deceived and seared. Relativism's effect has consequences when finally played out on its brutal stage of aggrandizement. Following an old African proverb, *the higher in the tree the monkey climbs, the more its nakedness is exposed.* Its ugliness is not always easily seen at first.

The Nazis in Germany fell into relativistic thinking, justifying the myriad atrocities committed against other people. In the end, the laws of human nature limped back onto the scene, expressing themselves through those that judged them in the aftermath of World War II.

Ask those cheated, lied to, or stolen from, or the families of those murdered whose lives were turned upside down. They will give an answer to those who believe they can function outside of the laws of human nature.

John Lennon recognized the law of human behavior written on our hearts in his song, *Imagine.* Imagine, he seems to say, if we all lived by the creed created in us. Imagine if we all behaved in a way that all mankind knows to be right. It's a wonderful song that brings warm feelings to the soul.

Unfortunately, he negates the end result by deceiving us in the beginning of the song that we can imagine away the forces of hell that war against us, and the heavenly forces in place to save us. Our own lives, as well as history itself, are laboratories of experiments that prove that hell on Earth cannot be simply imagined away.

Sure, we might imagine away strife among people, but what about floods, famine, diseases, landslides, and tsunamis that devastate people's lives? What about the problems you suffer to overcome? Or what about the struggles of someone you know? The song seems to be written for people who have never experienced a physical difficulty through a material agent. Or should we imagine away all that also?

We all have a heavenly potential in Jesus Christ to conquer hell's forces. However, to discover that potential, we mustn't try to imagine away the suffering of this world. When children are small, they may plug their ears, close their eyes, and loudly holler, "La-la-la-la-la!" Even they, however, don't completely believe that *that* will cure their problem. So what will?

The answer always leads us back to Jesus Christ. A question correctly posed is half the answer. Ask yourself, *what do I really want out of life?* Would you in the deepest part of your heart be satisfied if all your suffering, sorrows or injustices were to disappear and yet your life remain unfulfilled? Jesus Christ offers the only adequate answer. Even if it doesn't make your problems disappear, His answer will fulfill your life.

God created us to be free that we might freely come to Him. If He had eliminated all evil so that we had no choice but to do good, and then created us to love Him, that love would not have been freely obtained on God's part or willingly given on our part.

Many people today feel unloved. Perhaps you're one of them. How many of us, though, would honestly say we would want the love of someone who had no choice but to love us? How many people would want to force someone into a position where they possessed no option but to love us? It could never be true love given to us because it could never be willingly given by the other person.

Such love would never plumb the depth of that which our hearts desire. It would not be willing because our power to force one to love us would be greater than their ability to resist. We'd essentially become a cosmic bully in the possession of an unsatisfactory love.

So, we can see, the greatest and deepest love ever offered by a human is the love that can turn away from us. To have the abandoned love of someone who chooses to love is far more desirable than the love of someone who has no choice but to love us because we forced them into that state of being.

This is a much greater truth with God. God wants our true unforced love and allegiance. That's why He allows the world to exist with all the things opposite of His nature. All the evils of the world give humanity a choice.

It is because of His great love for mankind that God chose to leave the majesty and glory of eternal heavenly peace, joy, and love, and become a man who died a bitter death on a Roman cross. When John Lennon imagined away heaven, he also imagined away unfathomed eternal love. In doing so, he eliminated the real potential in the rest of his song.

The real potential for mankind is eternal love generated from outside of the human experience. I have learned to live with the suffering of being confined to a wheelchair for that kind of love. That kind of love I'll die for. That kind of love brings hope and courage to humanity beyond what John Lennon might have ever imagined. Mighty men have accomplished great deeds for that kind of love. Greater yet, though, for that kind of love, great men and women have led a life of joy and peace, hoping to point someone else to the same source of eternal love. There's the choice. The love of God or what we might imagine as we *la-la-la* our life away.

The power over evil lies in the gift of Jesus Christ's substitutionary death. It does not lie in your imagination. Jesus's work on the cross reunites you with God if you accept it and choose to trust and obey Him. Making that choice does not grant a life of ease. Nor does it grant a life of safety and security, fulfilling the ideals of the great American dream. It does mean that perhaps for the first time in your life, you'll discover something worth risking your life to pursue.

CHAPTER EIGHT:
What Does This Relationship with God Offer Me?

Every human being has a great privilege in that we can call out to God, especially in our times of need when the injustices, pain and sorrows in our life seem too much to bear. We have this privilege because Jesus Christ died a torturous death for us on a cross. We have an assurance that this is so because of His resurrection from the grave. But think of the enormous love of God that even allows us to have such a privilege.

Jesus became our great sacrifice because in His love, it was the only way to turn us to Him without forcing us to love Him. Our sins wage a war against

God. In love, God wages a war against sin through Christ's sacrifice to make Him more attractive to us than sin.

The greatest of warriors once drove a stake in the ground and tethered themselves between their home and their enemies. This proved their commitment that they would not abandon their loved ones to the enemy. Those who wished to be saved stood behind the great warrior.

When Jesus died, He drove a crude Roman cross in the cosmic soil of eternity. Jesus staked His life on the fact that He alone loved us more than anything else in the universe; He alone could save and bring us close to God. Moreover, though, He came back to life after being dead to prove He held the victory over death, mankind's worst enemy, and therefore no other enemy could stand against Him. The cross still stands as a testimony throughout history that Jesus Christ staked His life in defense of humanity.

People may scoff at this. They may deny it happened. They may spit on the cross. But they'll never make it go away. He still offers to stands in the gap between you and the injustices, sorrows, and sufferings in your life. He holds out nail-pierced hands and calls to you personally,

> "Are you tired? Worn out? Burned out on religion? Come to me. Get away with me and you'll recover your life. I'll show you how to take a real rest. Walk with me and work with me—watch how I do it.
>
> "*Learn the unforced rhythms of grace.*
>
> "I won't lay anything heavy or ill-fitting on you. Keep company with me and you'll learn to live freely and lightly."[1]

Will you accept that kind of love?

Chapter Nine:
Fear and Doubt

Events in life that are devastating to us often paralyze the mind. Emotions perhaps once unfamiliar to us now wash over life like tsunami waves. They pound and smother us. Uncertainties bringing fear, doubt, and despair keep us fixated, unable to move.

I once crested a ridge overlooking the Hoback River in the heart of the Wyoming Range. Deep winter had long settled over the landscape. A vault of thick ice entombed the river. Silence reigned. Below I spotted movement on the frozen waterway.

A moose calf stood splay-legged, head drooped in fear and despair. Huge puffs of its breath drifted in the morning air. The snow, churned up in a wide arch around its feet, told of a struggle underway. Then I saw something shoot from the surrounding trees, huge and massive.

This large coyote or perhaps a wolf began darting in and out skillfully, steadily sapping the moose's strength in small increments. Once the predator realized that moment of ultimate weakness, it would go in for the kill. The scene before me offered a window into an intricate dance between the struggle to live and death's exhausting process of stealing life.

Have you been there? Have you ever stood swaying in an intricate dance between life and death? Like the moose, have you ever been splayed out frozen in fear, doubt or despair, believing that any moment you'll cave in and become devoured?

In Isaiah 41:10, God said to Israel in their darkest moment when the Babylonian empire was about to devour them,

> "Fear not, for I am with you; Be not dismayed, for I am your God. I will strengthen you, Yes, I will help you, I will uphold you with My righteous right hand."

"JESUS CHRIST *cares for* YOU."

In those moments of utter dark despair in your life when it feels like your prayers bounce off of the ceiling, He gives you a personal assurance, "I am with you." When human weakness from fear or despair exhausts itself, He offers strength. To receive this strength, you need only to use the three weapons He provides for you.

God did not bring fear into this world. He offers power and love and a sound mind. To receive this, we must first humble ourselves and hunger for His strength. David prayed in Psalms 42:1, "As the deer pants for the water brooks, So pants my soul for You,

O God." We receive this strength when we use the weapon of prayer.

Satan is the author of fear and despair. He would have us blame our anxiety on God. By blaming God, we cut off our only true avenue of strength in time of need and fall into his trap. Remember Satan desires only to kill, steal and destroy your life. The last thing he wants is for us to pray.

Another weapon we have is what the Bible calls the sword of the Spirit. This sword cuts through the thickest darkness in our lives. Oftentimes people think that they can't ever understand the Bible when they read it. These are words of deception.

Throughout the halls of time, deception has echoed doubt over and again within the human ear. It has held mankind in bondage to the devastating situations running rampant in each of our lives. Yet we need only to pray that God will unlock the mysteries of His written word in our hearts.

Jesus said,

> "Peace I leave with you, My peace I give to you; not as the world gives do I give to you. Let not your heart be troubled, neither let it be afraid." —John 14:27

Has fear and despair stolen your peace, leaving you splay-legged and afraid of the things in your life that would eat you alive? Turn to God's word.

Finally, we possess the battle cry of praise. Darkness cannot stand in the light of praising God. Praise brings joy despite sorrow. It points away from fear and trouble. Praising God requires that we sacrifice those things to God. This may be the most difficult weapon to use because that voice of deception will object, "That sounds stupid!" Many battles have been won, though, throughout history by warriors singing their battle cries of praise against their enemies.

These are easy weapons to use but difficult to wield against the invading circumstances that slowly ebb strength. They may not relieve physical problems.

However, by using these weapons in the strength God offers, you'll overcome the horrible burdens that trouble your soul.

To overcome brings victory.

CHAPTER TEN:
Summary

Jesus said,

> "Take My yoke upon you and learn from Me, for I am gentle and lowly in heart and you will find rest for your souls. For My yoke is easy and My burden is light."" —Matthew 11:29

Jesus knew animals. He knew that sturdy and willful bulls could be raised up to be powerful beasts of burden if they would recon to their master's call. He knew that the strongest animal, tamed under the control of a man of great and gentle skill, could accomplish unimaginable work and not become weary. Jesus knew that it was the will—not the brute strength—that made the difference. Furthermore, Jesus knows your abilities.

A rancher that I know had a team of Percheron horses he wanted to train to pull his hay wagon. These high-spirited animals were not accustomed to pulling a load. The wagon following close behind these young, inexperienced animals left them jittery. Skilled in the art of training horses, however, my friend slowly worked them over time into the feel of the empty wagon, allowing them to become familiar with it before putting them to work. As days passed, he guided them onto a dirt road, traveling through the ranch.

Turning off the road, he headed them for a wooden bridge that crossed an irrigation ditch. As the wagon turned, it rocked forward a little, spooking the horses. They lunged to the left, missing the bridge in a panic, pulling the wagon down into the ditch. The stabilizing tongue connecting the front and rear axles snapped close to the front axle, dropping against the ditch bank. Unaware, the rancher, still holding complete control of the horses, tried to ease them beyond the ditch.

Confident these horses were strong enough to pull the wagon from the ditch, he gently encouraged them forward. As they powerfully shouldered the harnesses, the horses bent low to the ground, digging their feet into the soft, damp ground, but eventually they couldn't pull the wagon clear of the ditch. Upon examining the situation, he discovered the broken tongue. He knew the potential power of these horses, but even he was amazed when he discovered that those horses had driven the tongue into the ground six feet before they could no longer move.

Jesus knew the strength of a harnessed animal. Whether it was a horse or ox, He knew the potential strength they possessed. If the animal, whether it be weak or of great strength, allowed a gentle master to guide and control its life, then it could accomplish great tasks, regardless of its load.

“JESUS *knows* *our potential* STRENGTH”

At some point in our life, we will all struggle under the burden of some injustice, pain or sorrow. These will be as yokes with rough, sharp, splintered edges that ostensibly harness our lives and make us feel like they will tear us apart.

When Jesus said take my yoke upon you, He realized this. But His yoke is easy. His has the padding of peace and grace that passes all understanding that keeps our heart and mind on Him despite the heavy load placed upon us. Because of His love for us, He will not allow it to cut into our lives to steal, kill, and destroy our greatest potential.

Furthermore, He won’t add to the burden in our lives. Jesus knows and understands we already have enough to carry when we’re yoked to circumstances beyond our control. Many people want to avoid religion because of the added burden they believe it'll place on their lives.

Jesus doesn't offer you religion, though. He offers you Himself. Jesus offers grace. He offers friendship, power, and courage to get beyond your difficulties in life. Jesus Christ offers you the unforced rhythm of grace that you might accomplish mighty feats for His kingdom.

Are you weak hearted? He'll take you. Are you strong willed? He'll take you also. How about those dark sins in your life that cause shame that only you know about? Nothing is too great for Jesus to handle if you confess your sins and turn your lives over to Him.

He looks for those men and women, girls and boys, maimed and broken lives, weak or strong, bullheaded or sheepish, wicked or degenerate. All He wants is our willingness to accept His love and to allow Him to lead us for His sake.

Think about this. Most of us have heard the Christmas story. If you haven't, pick up a Bible and read about it in Chapter 2 of Luke, one of many books that make up the Bible.

God didn't make the road to Bethlehem smooth for Mary and Joseph. They had to travel to pay their taxes at the worst time for Mary's pregnancy. He didn't bail them out of this tax burden, either.

God didn't even supernaturally intervene to hold that *one last* room for these weary travelers. Poor Mary. She had to travel on a stinking, smelly donkey at the end of her last trimester. Imagine her discomfort as this animal plodded along. When she couldn't handle it anymore, she had to walk the uneven surface of a Palestinian road in sandals. And then the final straw—no room in the inn!

You would think by human standards that God would have at least *had their backs.* No, none of these things. He didn't even set straight the wagging tongues back home. You know, the gossip about

Mary being pregnant before she and Joseph were married? You'd think God would have at least done some of these things for Himself! After all, God became a man in these circumstances.

What *did* God do? Was He there? Did He care? Why *didn't* He at least come through for Himself? Well, He did. He gave Mary and Joseph the courage and strength to overcome their difficulties and the hope and faith to endure the hardships. Jesus was born in a dirty stable, so He would experience the same dirt and hardships we've experienced in life. He loves us so much that He wanted us to know that He knows firsthand the misfortunes we suffer.

He's not a distant God watching you from a cosmic telescope, waiting for you to screw up. In Jesus Christ, He's seen misery close up. He knows temptation. He's fought off Satan's attacks.

God works from a bigger plan, in which He graciously invites us to participate. It's a kingdom plan. It's one He started in the dirt of this planet we call Earth. Don't miss this. It's a plan to eliminate sin and evil permanently, not just make it tolerable or acceptable to our human liking. God hates sin.

Do you know what this means for you? It means that He cares a whole lot for you! Really. He cares personally. Still don't believe me after all these pages? Get on your knees and ask Him. He'll let you know.

To those who will take up His offer, He'll give padding to their yoke. To those seeking Him, He'll add the unimagined trait of courage and hope that only God can give. To those who are defeated without hope and ready to end it all, He'll give strength to live a life worth living. Most of all, Jesus offers the only true eternal love you can bet your life upon.

Jesus did all this for me. My empty, broken, hopeless life lay in ruin until I gave it all to Him. Jesus Christ padded my yoke and made it easy for me to carry.

So, I offer this very specific message to the one reading this right now because Jesus can fill you with the confidence and strength found only in God. My friend, I hope you too might find the saving love of Jesus Christ, despite the heavy burdens you carry.

Before you're able to live to your fullest potential in which God created you, you must come to grips with this simple question: Who do you say Jesus Christ is?

God is not constrained by your circumstances. Through your obstacles, God can create vehicles that will accomplish great service. Through your weakness, God can deliver great strength. In that strength, you will receive the power to overcome.

Through your life, God has a plan of welfare, not calamity, to bring you a future and a hope.

Do you settle for far less than what God has planned for you?

End Note

Scripture quotation on page 81 from *The Message,* Copyright © Eugene H. Peterson 1993, 1994, 1995, 1996, 2000, 2001, 2002 Used by permission of NavPress Publishing Group (emphasis added)

Made in the USA
Monee, IL
08 November 2026

67944872R00069